Women and I.T. Careers

Why Women Are Leaving the Ranks of I.T. Careers and Why It's So Important They Stay

Dr. Cedric Alford

ISBN 13: 978-0-9971448-0-2
ISBN 10: 0-9971448-0-7

Printed in the United States of America

TABLE OF CONTENTS

INTRODUCTION

A s a 20+ year veteran working in the technology industry, and as a witness to its explosion of growth along with more and more women entering the work force, I wondered why more women weren't entering the IT field.

After all, technology was advancing at a rapid rate and offered exceptionally good paying jobs with a secure future. The tech industry in Silicon Valley was a hotbed for innovation, with other cities such as New York, Seattle, Chicago, Miami and Austin emerging as tech hubs. In fact, the Lone Star State of Texas, whose burgeoning tech scene is drawing IT professionals from other cities around the world to relocate there, is poised to roll out tens of thousands of tech jobs by the year 2017.

Women are in a prime position to take advantage of these new jobs. However, statistics show that women are not growing in the number of IT professionals, and women are also not appearing in IT leadership roles. The importance of why fewer women than men chose the IT field was so prevalent for me to understand why that I based my doctoral dissertation on the subject. I want to share my findings of researching 10 women, all IT professionals, in the Dallas/Fort Worth area who resided in leadership roles - project manager or above - for a minimum of at least three years in those higher positions. By using a smaller number of participants, this allowed for a more effective approach, bringing out the individuals' personal experiences and perspectives.

In all of my years of experience, I have probably consulted with around 400 of the *Fortune 500* and *1000* companies. During this process, I have rarely seen women in the "C" Level conversations I have had, but saw many on the way to the office of the Chief Executive Officer, the Chief Marketing Officer or Chief Operating Officer. Not only did I take on the opportunity to understand this from a research standpoint, but over the course of my 22 years in IT, I have started companies from the ground up or grew existing technology teams by adding women to previously male-only teams.

It is my greatest hope that the findings of my study encourages others to develop their own thoughts and to create additional knowledge to advance understanding in this area, and bring back women to tech jobs as a global solution to serve the world's needs more productively and with more equality overall.

CHAPTER ONE

The Paradox

I t is an extremely somber fact that women are under-represented in the IT industry in America. For that reason, this implication will have a substantial impact on our overall global competitiveness. As for the IT industry, it will affect our ability to foster innovations in science and technology.

This notion is not just an opinion I hold as a technology executive. Bill Gates, the creator of Microsoft, was speaking in Saudi Arabia to a group of IT professionals. He noticed that on one side of the room, all of the men sat together, representing four-fifths of the number of people in the room. The remaining one-fifth of the room was comprised of women, all cloaked in black with veils covering their faces.

During the question and answer session, one audience member said that Saudi Arabia aimed to become one of the top 10 countries in the world for technology. "Well," Gates answered, "if you're not fully utilizing half the talent in the country, you're not going to get too close to the Top 10." After hearing his response, the women broke out cheering.

The thought-provoking statement Gates made can be applied here in America. IT is the fastest growing sector of employment in this country. In 2007, the National Center for Women and Information Technology (NCWIT), an organization that ensures women are represented in the IT industry, reported in its annual scorecard that over one million IT jobs will be added to the U.S. workforce by the year 2014. Upon writing this book in late 2015, we see that this earlier projection came true. The jobs are there. Yet, women face many challenges unique to them to get these jobs.

And, out of those one million jobs, how many women will hold leadership positions as well?

Some women will advance on and obtain IT leadership roles, but the vast majority of those women will not. Why? The answer to that is plain and simple: Bias still exists in the mostly male-dominated vocation of the IT industry. More disheartening is that the ideals people have of leaders and managers are quite different from the ideals they hold of women.

I'm not saying that all IT management teams see women as incompetent or lesser than men, however, barriers to career progression and their ability-to-achievement gaps have resulted in women leaving IT altogether. While some organizations have implemented various programs to recruit more women, only marginal success has been seen in the percentage of women obtaining IT positions, let alone those that progressed into a leadership role.

Furthermore, the required skill sets and perceived barriers common to women in IT leadership have not been a high priority in research, so we don't know much about it outside the confines of the corporate IT walls or what we've witnessed on our own, which is another reason why I chose to study it at an academic level.

The Barriers Women Face

Under-representation of women in IT leadership roles is a well-documented phenomenon, with women only experiencing small growth in this field. In investigating the root causes of women in IT not achieving the ranks of management at a level that is equal to the percentage of women in other leadership roles, three categories of barriers to career growth for women in

IT were found:
1. Education aspects and family
2. Corporate cultures
3. How well they related to others

In the next few chapters, we'll identify types of organizations and reasons for these biases and provide some background information on organizations and how they think and work before getting into the actual individual summations from my research, but here are some statistics to justify these findings.

Despite an increase in the number of female graduates in technology programs and an overall representation of over 50% in the United States' professional workforce, the U.S. Department of Labor reports that only 22 percent of women work in technology fields.

Of these women, only 12.4% are on the board of directors at *Fortune 500* companies, with less than 1% achieving

CEO positions. Even in non-management positions, the Association of Women in Science reported that although women make up more than 29% of all full-time technology jobs, women in technology positions earn only 90% of the salaries of men.

Furthermore, women endure a high level of scrutiny when they do obtain a leadership role. When women are offered leadership roles, it's quite possible those jobs are in areas that could be considered more precarious than those occupied by men, leading to positions that have more built-in failure potential rates than those coveted by men.

And in spite of what they profess, gender issues do seem to weigh heavily for corporations. For those women that eventually achieved leadership roles within an organization, one colleague of mine said that "it is not really known whether the female directors have been appointed due to their skills, the demand of the organization, or if they have been appointed because of the current external pressure of gender equality."

Of these 10 women who worked with me on this study, I was quite surprised when talking to those who were undergoing performance reviews that they generally found out that managers attributed their success to "luck or lack of task difficulty" rather than acquired skill.

Luck? Such as in placing your entire education and career on a red or black bet at the roulette table? A piece-of-cake job? Not skills?

This revelation seemed peculiar to me, especially coming from managers at the top of their game. Workplace attitudes has a lot to do with this. Gender does have an impact on leadership potential within organizations, as men are seen as natural leaders and women are not.

Workplace culture is another factor in male-dominated organizations where women seek leadership positions.

In fact, the corporate culture within an organization may be the key contributor to a woman's ability to obtain a leadership position. We could also go back and say that women seeking leadership roles in IT may be at a disadvantage because of earlier attempts to socialize them away from mathematics and science when they were young by family members, peers and even educators.

For women seeking management positions, family demands, bias in performance evaluations, favoritism towards white males, and assignment to female-oriented tasks and markets are also possible contributors to the low numbers of women in IT management.

Women also face challenges in obtaining IT leadership roles because of contradictions between the nurturing nature of women and the anti-social, male-dominated culture of IT. When faced with the challenges of overcoming the barriers to advancement in the organization, some women choose to leave the organization or seek a different career path.

For women who decide to persist and take on the challenge of pursuing a leadership role in IT, many encounter the social organizational barriers of career advancement, career persistence and career choice that are engrained in the culture of the organization.

The struggles of women in organizations show that some women in leadership roles have challenges navigating the unrealistic expectations that lead to negative performance reviews and eventual involuntary separation from the organization. Performance reviews definitely contribute to women having higher attrition rates in IT than men.

The only way to overcome attrition and increase the number of women in IT leadership roles is to increase the number of women in IT in general.

Women in IT perform poorly on performance reviews because they are too few in number in the male-dominated IT industry.

The lack of role models and low percentages of women in the ranks of executive management are also factors impacting the advancement of women in organizations. This argument can be supported by stating that individuals who have stronger identity, competence and growth opportunities thrive when they are coupled with others that share similar racial, gender and social traits.

Training and mentoring are viewed by women as key contributors to their success in obtaining an IT leadership role. In the absence of *homophily,* where individuals with like interests, gender, group or ethnic affiliations tend to interact in a mutually-supportive manner with one another, it can negatively impact the ability of minority groups (women) to gain any significant support within the organization.

From the information gathered, the potential causes of the under-representation of women in IT could be due to unsupportive educational environments, the lack of mentors, obstructive social norms, unrealistic and stereotypical expectations of women in the workforce, and because women are perceived to be less able to perform technical tasks than men.

And while both men and women have to overcome the same barriers for leadership progression, the barriers for women are much higher and problematic than those experienced by their male counterparts. These are cold hard facts supported by evidence, not opinions.

CHAPTER TWO

Organizational Structures

Although the central research question guiding my study was, *How do women in IT leadership roles perceive how they obtained and sustained their positions*, I used three different approaches to gain more insight into this matter.

First, a company's organizational culture can provide visibility into the collective psychology of the organization. Also, how the thoughts and practices of the members evolve into the culture can be seen.

This provides a blueprint of the values, beliefs, language and actions of the organization.

Taking an Organization into Consideration

Organizations have been around for centuries and play an important role in society. They are generally shaped by broader processes driven by social and cultural forms.

I'm including this rundown into the different types of organizations to provide a better picture for you into the IT company you're planning to work for. Your career depends upon recognizing this facet of the business, and will help you determine whether you're a strong fit within the organization.

****

Organizations can be described by using one (or more) of the following five models:

1. Rational
2. Natural
3. Open

4. Socio-technical
5. Post-modern

Rational organizations consist of formal structures, roles and processes regardless of the size and scope of the organization.

Natural organizations take into account the social interests of the individuals within the organization and are based on the premise that no organization can devote all of its focus on producing products and services. It must be treated as a system, adaptive and considerate of human values and motivation.

Open systems are self-regulating and bring in capability from the external outside environment. It exchanges material, energy, people, capital and/or information with its environment.

Socio-technical systems show the social aspects of change, including social and technical systems, and the concept of innovation. They are dynamic and operational, which therefore supports the interaction between human beings and technology.

Postmodernism systems are a threat to traditional natural, open and socio-technical systems, as they further de-emphasize the role of the human resource in the organization.

Organizational Culture

While people are working, they are not just producing goods and services, paychecks and careers, they are also producing culture, i.e., a sub-culture.

Organizational culture can be redefined when the boundaries of an organization extend to contractors and other external partners.

Cultures can be both inclusive and exclusive. While organizations profess to the world their value of diversity and a multicultural workforce, and in spite of laws to protect discrimination against women and minorities, organizations systematically work to block the transfer of sub-cultures into the fabric of the organization.

Gender plays a role in organizational sub-cultures from the standpoint of gender-related behavior. Organizations, and groups within organizations with great gender diversity, are more open to nontraditional gender behavior versus those that have a dominant gender, such as men. Both men and women in job positions held by a dominant opposite sex can have decidedly different outcomes on productivity. Organizational culture can also drive language-based gender expectations.

The three types of languages we see in organizations are:

1. Military
2. Sports
3. Sexual language

In each case, the languages that had masculine vibes more often than not had left out women from the base language of the organization. This results in *exclusion*. And most people would agree that the male culture and language preferences dominate most. While women are rewarded for showing masculine leadership and behaviors, men are criticized for acting in a feminine manner.

Organizational cultures such as these create harmony for most males and disharmony for most females. One study that followed the profiles of 50 successful women found that the most successful women were "politically seasoned" and able to navigate both masculine and feminine expectations of performance and behavior within the framework of the organization. The difference between those who made it to the top of their vocation and those who do not appeared to be solely based on *political skill*. In organizations with an ingrained political culture, women generally faced resistance from men when women attempted to develop political relationships within the organization. In order to combat this, the organizational structure and process must provide women with opportunities for the development of the skills necessary for collaboration and diverse networks.

Unfortunately, each organization that was studied proved to be very different, the undervaluing and underrating of women appeared to be a common theme. In many cases, gender had been embedded as part of its culture.

Take this example about one remarkable woman I encountered. Back in 2009, I was recruited by a company to build a marketing automation practice to go alongside its existing and web analytics practice. Already having a strategic relationship with major players in the marketing automation space, the organization was seeking expertise to take advantage of a new relationship with the then leader in the segment.

I accepted the role as Vice President of Sales and Marketing Automation and immediately jumped in to begin building the team and relationships with the partner. Because the organization was predominantly remote, I did not immediately notice that all full-time employees were male, and the few females at the company were each contractors. This became clear at the bi-annual off-site that was only for permanent employees.

To say the least, this experience was a bit awkward given my passion and research interest in diversity of technology organizations. On the plane ride from the off-site, I realized that I had a great opportunity to change the demographics of the organizations through my next few hires.

After getting back to the office, I pushed back on the profile of the resumes that were previously provided by our recruiting arm. I asked for the priority to be first placed on identifying a greater diversity in talent; in particular, ensuring that I had an equal number of female candidates to pursue as I did males. While it was only logical to hire the best available candidates, I was pleasantly surprised to see that the best available happened to be three females – one a developer, one a project manager and another that had actually been trained and deployed by our partner organization.

One person, specifically, Debbie, was hired on the spot. She not only had an impeccable record of performance and project leadership, but she seemed quite motivated and eager to hit the ground running.

In a typical engagement, I would work with the team to kick off the project, manage executive interviews and determine the optimal path towards a solution. Often times, I had to assign multiple consultants, a project manager and a technical architect to ensure the success of the engagement. However, with Debbie, she was able to manage each of the functions without much support.

Initially, I saw Debbie as a rock star as well as an example at the male-dominated company that it was okay to trust strategic relationships to women. However, over the next few months, I noticed that I began to get emails critical of Debbie—her communication style as well as her ability to manage her projects. My operations manager noticed that Debbie was beginning to work the majority of the hours on the engagement versus delegating the tasks to team members.

It should be noted that none of this feedback was coming from customers, as one customer stated on a call, "Cedric, you'd better be careful. Debbie will be on our team next year!"

No, this feedback was internal. Being a mentor to Debbie, I brought this information to her attention during one of our weekly connects. I explained to Debbie that my style is to be transparent, and if positive or negative feedback comes in, I will discuss regardless.

I asked Debbie her take on the feedback and she shared that each decision she makes is second-guessed. The male team members interrupted or questioned her decisions in front of customers on calls. She explained that she was frustrated because for whatever reason, when she had an internal call, there appeared to be a festive atmosphere as she joined the call, but immediately after she spoke, silence or tension ensued. Her biggest frustration was the fear that her customers would now begin to also question her approach to their project because of the treatment she received from fellow team members.

She stated that the reason she had been working the majority of the hours was because she could no longer trust the team members. To get over the issue, she began taking them off her projects and working longer hours to cover the entire engagement. As I listened to Debbie, it became clear that in my organization, the men appeared to also be falling directly in line with research in that Debbie appeared to be judged harshly, was seen as outside the dominant group and had her communication skills questioned. While she was terribly talented, there appeared to be a separate set of criteria for Debbie than what was used even in the most basic performance assessments of her male counterparts.

Being a small company, I was one of three executives that included another vice president and the president. Interestingly, after discussing the issue with the executive team, it was pretty clear that the issue was perceived to be with Debbie versus the culture and sub-culture of the organization which catered to men versus women. That translated into the males in the organization who felt comfortable "giving women a hard time" because they understood that there would be no consequences at the highest levels of the organization.

Over the course of the next year, I attended a number of internal calls with Debbie and the internal team as a means of minimizing the tension. I even added "Senior" in the title of Debbie as a means of showing the other team members the opportunity to move up in the organization when you deliver high customer satisfaction.

However, Debbie's performance began to significantly decrease as a result of the fact that she resisted pulling in other team members to support her projects. The customers loved her, but her team members hated her. The unfortunate reality of this situation is that in a very short amount of time, the male-dominated organization transformed a vibrant, intelligent and motivated female leader into an angry, paranoid and ineffective follower. She chose to not have a team. She chose to become an individual contributor in a role that required great leadership.

Yes, she had great skills and a bright future, but it was very unfortunate when I was unable to get her to turn the corner back to the individual I initially hired. The damage had been done and the opportunity for her to lead in my organization was lost. This situation served as yet another indicator of the need for more organizational awareness of the impact of culture on women in mostly male-dominated organizations.

The Glass Ceiling

Let's examine the exclusion of women from the higher levels of employment. My studies found both informal and formal barriers existed. Certain groups monopolize advantages by closing opportunities to inferior or ineligible groups. The gender role perspectives of both men and women directly impacts their life experiences. As each group creates preferences and expectations of gender in life, men and women take their perceptions and attitudes into the workforce, contributing to its sub-culture. As women enter the job market, women mostly concentrate in "jobs associated with low levels of prestige or pay". As women develop goals for leadership, they find that the perception of their traditional role in the workplace supersedes their qualifications and thus reduces their promotional opportunities into top management positions. This phenomenon, known as the *glass ceiling*, presents an invisible barrier for women to achieve leadership roles.

It could be that the key difference between women who achieve top leadership positions and those that do not is political skill. Career development of women is different from men and requires additional support for women to achieve comparable levels of success. Women find that domestic responsibility may be a barrier to leadership attainment within an organization.

Also, it can be seen that gender issues existed by stating that far fewer women in management positions had families with children when compared to their male counterparts.

Women continue to be limited from attaining leadership roles by the invisible glass ceiling, the invisible, gender-based barrier for women achieving the highest level leadership positions in an organization. Sex segregation of jobs points to the work and family conflicts of women. Therefore, now it's more important than ever to research the company before you get hired.

Gender bias most often begins in the interview process and extends to post-employment performance reviews, where males tend to obtain more favorable reviews than women. Although women are beginning to break the barriers to leadership roles in some industries, scrutiny in their performance by their managers is much higher than that of their male counterparts. Often times this points to the relationship between gender and social desirability. Women have been known to frequently over-report favorable behavior on surveys when asked questions about available resources, fairness at work, care and concern, employee trust and company reputation. This suggests that women are more conscious of the impact of a negative rating and tend to compensate by scoring themselves higher as a group than men. It is possible that women provide more favorable personal responses as a group because of what they perceive as social consensus and a low level of consequence for a misleading answer.

An important note though is that the progress women make concerning the glass ceiling may be misleading. Women are being offered leadership roles in areas that may be more precarious than those occupied by men, leading into premature involuntary or voluntary separation from the company.

Additionally, the continuous process of scrutiny that women state they endure on individual performance evaluations when they achieve leadership roles leads to another phenomenon called the *glass cliff*.

The Glass Cliff

The glass cliff points to a phenomenon where women are promoted into roles that carry high risk and high visibility, but end up leaving under involuntary circumstances as a result of the impossible expectations of success in their leadership role.

While women are under-represented in attractive leadership positions, they are over-represented in those that could be labeled as risky. When women fall off the glass cliff, it reinforces in society that women aren't able to be good leaders.

Some recent examples of this can be seen in the exits of Ellen Pao, the CEO of Reddit, Carly Fiorina, CEO of Hewlett-Packard, or Mary Barra, appointed CEO of Sunoco just as its market shares took a 52-percent dive. The list goes on.

The Concrete Ceiling

The concrete ceiling – denser and less easy to shatter – not only restricts access to top-level positions, but middle management positions as well. It applies primarily to women's issues, but evidence shows that minorities are faced with insurmountable barriers as they attempt to move upward in their career choices.

Derailment

Women are frequently denied training and promotion opportunities due to the perception that the family needs of women will take priority over the time requirements and requests of the organization. As women begin to hear these messages repeated, it destroys their confidence in all but the strongest or most militant, so they stop competing against their male counterparts.

This amounts to derailment, which occurs when individuals either do not reach their potential or leave the company prematurely or involuntarily as a result of having moderate skills in one area, either leadership or management, with low skills in the other areas.

There is no denying that challenges in pursuing career paths because of family or society expectations may discourage some women from pursuing certain careers, regardless of their prerequisite education and experience. In a study of men that did not have a spouse in the labor force, it was found that those men achieved greater success than men or women whose spouses work. In other circumstances, men with a spouse who did not work had greater organizational success than both men and women whose spouses worked.

Women who obtain and sustain their leadership positions must take into account the organizational environment where they are looking for a leadership role.

Thus, a woman's success can possibly be at the cost of a traditional family life. Long hours at the office, extended periods of travel, overnight meetings in different cities, dinners to entertain clients - all take you away from a normal home life where sitting down for dinner at the stereotypical

"5:00 pm hour with the family" is rarely possible. This thinking is outdated anyway. In the last 60 years, the role of a woman remaining at home in a family care role is the exception today rather than the norm.

Society tends to hold women to a set amount of feminine standards, although "femininity" is a term that has negative meaning. The dominant male images bring out a negative view of feminism or "femaleness", thus leading to barriers being presented to women seeking career advancement through established male paths to leadership. This creates a paradox for women in role expectations - conforming to femaleness and complying with male-oriented work practices.

Authority, Power and Influence

Authority, power and influence exist as part of the culture of an organization. *Power* based on authority is the ability to require or to get

34

followers to act to achieve a goal or purpose. *Authority* refers to the right to do, to create, to activate movement in others toward a particular aim or purpose and can include legal documents or delegation. Power and authority usually work together harmoniously.

Influence is a loosely defined ideal that is based on rhetoric, on appearances, on images, whereas authority is based on a preexistent knowledge shared by all. Power appears as a force, but authority without power is empty and meaningless. Power involves having the authority to produce an action in someone else and influence involves the less direct means of achieving the same outcome.

Gender also has an impact on power, authority and influence within an organization. Power and authority within organizations generally favor a gender group as a whole or a group of individuals within a gender group.

Organizational Fit and Gender

An individuals' perception of their fit into an organization can create negative emotions related to stress, job performance and their aspirations for management positions. Organizational fit is generally based on the compatibility between the goals of the person and the goals of the organization.

Gender, organizational structure and processes are related. Women who succeed in these organizations find themselves overcoming several barriers to obtain equal status with their male counterparts.

Three distinct groups of gender imbalance occur within organizations:

1. Where women were outnumbered by males at both the upper and lower level of the organization
2. Where women were in

abundance at the lower levels but low at higher levels

3. Where women were represented at all levels of management within the organization.

Two individuals can have the same positional power within an organization, but their success could be greatly limited by their scope of influence.

Gender imbalance within an organization creates a culture that is hostile to women, and compromises or excludes them from social networks dominated by men.

Gender mix is an important factor concerning the progression of women into leadership roles, which is why you should thoroughly look into an organization's history before accepting a position. It is also wise to talk to someone who works within the organization to get a feel for their culture.

CHAPTER THREE

Women in IT Leadership

Leadership is a complex process involving the leader, the followers and the situation. Wide acceptance of leadership traits plays a central role in how leaders are perceived, and organizations that define leaders by traits and not behaviors result in the continuing on with stereotypes that negatively impact women. Until recently, the study of women and leadership was largely dismissed, with leadership addressed more towards men.

Women in leadership is a topic that is often talked about from the point of view of sex and roles, focusing more on the traits and behaviors of women in leadership rather than applying it to women in leadership positions.

CEDRIC ALFORD

Thus, the path to leadership for men and women can be considerably different. Glass and concrete ceilings – less impenetrable as glass but sharing a similar meaning – highlight the challenges and blockages which women experience in pursuing promotions and remaining upwardly mobile. From the start, gender influences form about men and women in leadership positions despite actual behaviors. And sometimes, male attitudes turn negative when the female leader becomes a direct manager of the male.

Some organizations treat men as having natural leadership qualities and that women must develop skills in leadership. In these environments, women run the risk of entering an organization with a negative perception before they attain or are considered for a leadership position.

Higher Education and IT

Nevertheless, under these

particular circumstances, an increasing number of women have pursued careers in IT over the last decade. However, the increased number of female college graduates has not resulted in an increase in science, engineering and technology leadership positions for women. Less than 5% of all IT management positions are filled by women. Those who do seek technology roles generally find challenges integrating into the organizational culture and obtaining a position in management.

One challenge involves whether women choose less competitive careers as a result of their need to prioritize family over career. This challenge can only be addressed on an individual basis.

Other challenges exist because college ranks may steer women away from IT fields. The NCWIT also found evidence that women are more interested in using computers to solve problems rather than actually working on the hardware of computers.

> **Some women in IT leadership roles struggle with family demands, bias in their performance evaluations, favoritism towards white males, and assignment to female-oriented tasks and markets - all seen as key contributors to the lack of women in IT management.**

While working on a strategic marketing engagement for a major Consumer Packaged Goods (CPG) company, I had a pretty high level meeting with executives from all around the organization. The goal was to stop the bleeding that the company was experiencing in terms of market share by helping the organization automate its digital marketing strategies. While in the meeting, I noticed that only one of the individuals in the room was female, not including the five people from my team. So, in a strategic meeting with 22 individuals, only one person was female.

As the meeting progressed and I facilitated the discussion, I noticed that each time a question arose that required a bit more detail, the men deferred to the only woman in the room. This went on for a few hours before the meeting was done.

Immediately after the meeting, I saw the lone female on our way out of the building. I reached out to her to confirm her contact information and used the premise that I would like to pick her brain on some topics discussed at the meeting. A few days later, we talked over the phone. I asked the obvious question of, "Hey, you were not on the invite, but I noticed that you had deep knowledge of the issues and the systems as well as an idea of the plan to get us from Point A to Point B. What is your role and why have you not been included in other discussions?" Then, I told her, before you answer, let me just share with you the fact that I am both a technology executive and expert on women seeking and obtaining leadership roles in the technology industry. "So, if at all possible, let's be candid," I said.

She seemed a bit relieved with my approach to the discussion and offered thanks for noticing her "obvious" frustration. She stated that she was considered the "team leader" for the initiative but did not have the official title.

She went on to say that she had more education than the men, but because the men had relationships with each other and they had worked at the company longer, she could not find a way to stand out and speak up to formalize her role. She felt that what she did on a day-to-day basis was important, but she was only invited to the meeting because she had the answers and they did not.

It was only after I explained the depth of the discussion that her "peers" forwarded the invite. She also explained that after the meeting, she was disappointed to be assigned each of the follow-up action items, when the responsiblllty of those items were with the men. She rounded out the conversation to explain that she did not expect it to be so hard to find legitimacy in an organization when she consistently achieved whatever goals and objectives placed in front of her.

She also said that being the only female – and an African-America female – put her in a position where she knew that there was a problem, but that she had no one to go to for support.

As she shared her story, I couldn't help but to smile and have a heavy heart at the same time. The smile was due to the fact that her concerns, stated barriers and issues were direct validations of the findings of my research. The heavy heart was due to the fact that this was a gifted, smart and professional woman that was almost in tears speaking to me about the plight that she had to have in order to glean a "successful" career. Her emotion was very similar to that of many of the women that I studied, as one participant also tearfully shared her regrets of not having children because her manager at the time explained that pregnancy would destroy her career. The first individual finally came to terms with the reality of her situation and decided to change industries. She is now in education and finding fulfilment close to technology, but outside of the technology field.

When you consider the fact that her organization was global and she obviously held the knowledge to support their global transition that none of the men held, the implications of her leaving could be felt for years. Not to mention the other women in the organization that saw this happen and how the organization let the blue chip talent leave without a title or any indication that she mattered.

Along this vein, I was highly interested in one study that captured women holding IT leadership roles in the Washington, D.C. area and explored the perceptions of barriers and skills of women in executive positions at IT companies. This study discovered that women join IT organizations for different reasons that ranged from career opportunities to salaries to opportunities that presented challenges. The majority of participants held at least a master's degree in an area that was not technology related, and enjoyed being in "a minority". The barriers for those women included the "good ole boy" system as a major challenge for them, and that workplace culture influenced the success of obtaining leadership positions.

Only 5% of these women were motivated to pursue a career in IT as a result of their educational backgrounds. Twenty-five percent of the participants had a passion for technology. Most of them also transitioned into the IT industry from some other field. The study concluded that learning the culture and politics of an organization are important characteristics for women in seeking to obtain an IT leadership role.

Their most common experiences were to get along well with their peers, contribute to the bottom line, their own personal drive and motivation, on being a visionary for the company that they worked for, customer satisfaction reasons were given as well as networking capabilities and maintaining visibility within the organization.

CHAPTER FOUR

The Study

I chose to study 10 women who held leadership positions in the IT field in the Dallas/Ft. Worth, Texas area. Because there is so little data out there about this, the goal was to identify common perceptions and any shared behavior within the group to understand why women weren't pursuing or maintaining IT careers.

The qualitative phenomenological methodology in the assessment was chosen as a result of the central research question of *how women in IT leadership roles perceive their ability to obtain and sustain their positions*. This type of research strives to better understand the social situations, events, groups, roles or interactions of an individual – the true human experience. Complete anonymity was guaranteed to the women by the signing of confidentiality agreements and informed consent documents – all standard procedures for Internal Review Board (IRB) guidelines, with checks to ensure the validity of the responses. All interviews were conducted face-to-face.

In order to get a wide range of different viewpoints, the ages of the participants varied as follows:

1. 1 - 21-30 age group
2. 1 - 31-39 age group
3. 4 - 40-49 age group
4. 4 - 50 years or older

All participants had eight years or more of experience in IT, with seven having 15 years or more. Three participants had three to four years of IT leadership experience; two participants had 7-8 years of IT leadership experience, and the remaining five participants had 10 or more years in IT leadership experience. Their job title had to be one of the following: Chief Executive Officer, Chief Financial Officer, Chief Technical Officer,

Chief Information Officer, President, Vice President, Superintendent, Administrator, Commissioner, Director, Manager, Supervisor or a Project Manager. Nine out of 10 women graduated from college with non-technology degrees.

My particular study is unique in that it expanded upon research by adding the lived experiences of the research participants as additional data elements for improved understanding, and went beyond only IT companies to include women in IT leadership roles regardless of the industry because this experience is not unique to only IT companies. Out of the study, many important revelations occurred, with 10 major themes emerging.

The 10 Major Themes That Emerged from This Study

1. The majority said their influence to pursue IT as a career was based on exposure to the field while working in another industry.

Their initial career paths were in popular, non-technical fields at the time they chose a course of study. Three participants majored in computer science. Their entry into the job market as non-technology workers was more due to a lack of knowledge of IT during and after college than wanting low paying, yet prestigious jobs. Once out of college and into the workforce, the women changed their career paths as more information became available about the IT field.

Their access and exposure to IT while in the non-IT roles influenced their decision to pursue an IT position. The initial positions required work experience, but required no degree in technology. Despite exposure to IT after college for some of the women, all were able to able to obtain and sustain a leadership role for at least three years at their companies. Six women shared this perspective. None had fewer than 13 years of IT leadership experience.

Although the NCWIT findings claimed women are interested in using computers to solve problems rather than actually working on them (hardware), the women in this study did not support this. They were drawn to it because of the challenge.

2. The majority said they were not prepared by education for their initial jobs in the IT field.

Their responses confirmed previous findings on this matter. Seventy-percent of the study participants felt they were not prepared for their careers by education. Given that their first degree was not technology, it's clear that this was the most common perception among them.

One woman IT leader stated, "In terms of experience, I was very well equipped. The experience was there. They had no problem in trusting me to learn what I needed to learn, to do what they wanted me to do...it probably would have helped for me to have more direct education, but the majority of the other IT people have the same amount of education that I do." Only three women had a technology degree prior to obtaining their IT leadership role, yet all were able to sustain their careers for at least three years.

3. The majority said they did not take the ideal career path to their leadership roles.

With fewer than 5% of all of the IT management positions filled by women, one of the possible reasons for this phenomenon could be the indirect path women take in getting into the IT industry. Only 30% of the women held technology degrees and 80% revealed that they did not take the ideal career path to leadership, although each were able to obtain and sustain their leadership positions. All of the women indicated entry level experience as the ideal career path. One felt that women must possess credentials beyond what is included as the minimum requirement on a job description.

Given the educational and initial career profiles of the women in this study, the possibility exists that interest and experience are better indicators of success for women that obtain and sustain a leadership role in IT than work/life balance issues. It may be possible that the alternate career path the women took in entering the IT field actually made them more qualified by providing them with exposure to what was required of IT workers (in terms of time and commitment) prior to changing their career paths. Regardless, the women still chose the IT field despite the challenges they observed while in other jobs.

4. Half indicated that their career path was different from the path taken by men.

Half believed that their career path was different from the paths experienced by men in their respective organizations. One IT professional explained, "I can say that of the 10 men, I was the only woman when I came to the group. The men don't have a degree, per se, but I had to have one. For new hires, they hired them because they were technical installers...I don't feel like a woman would be able to have been taught to do this job."

Five women found ways to get around career path differences to pursue their leadership roles. Although eight of the 10 women surveyed had children, all but one indicated that their career choice presented work/life balance challenges. Regardless of the challenges, all felt that family was as important as their career, so balance was a necessity.

5. The majority said that the reason more men are promoted seems to be men are more comfortable working with other men.

The women believed that more men were promoted to leadership positions because men are more comfortable working with other men, i.e., homophily, or love of the same. One woman stated, "When I go into a meeting, I take them off their game. They're uncomfortable. I like dresses, I wear dresses. They say, 'I can't say that, we've got a lady in our presence.' I say, 'You've said it before, say it now.' They are not comfortable around me." Seventy-percent held this perception. Two felt that women were not promoted to leadership roles either due to male stereotypes or simply because men felt more comfortable around other men.

Although all of these women each held leadership roles at their organizations, additional self-promotion opportunities for them happened with the perception male colleagues held of them rather than the results they created. The most common way they managed to sustain their roles was through controlling their emotions and in developing coping mechanisms. This enabled them to coexist with male team members who they felt did not respect their skills.

Although the concept of femininity is complex and not clear at times, dominant male perceptions of femininity leads to obstacles presented to women who are seeking career advancement through established male paths to leadership. One woman illustrated this point. "Because it's a male-dominated society, I think that it's hard to get away from that. Same thing with race, they got people who are used to doing what they're used to doing...they're more comfortable with someone similar to themselves, so if a male is in the power role, which they typically are, then they're going to connect better with another male."

The experiences of these women are not uncommon or new. Women find themselves overcoming these obstacles to obtain equal status with males, however, the women studied here saw that the culture of their particular organization contributed to a *reduction* in barriers for them to achieve promotional opportunities for IT leadership. In other words, their organizations' philosophies were healthier towards women in powerful positions, therefore, it proved to be easier for these women to obtain leadership roles.

6. Half indicated that women are under-represented in IT because it is perceived by women as a male-dominated field.

The findings supported the NCWIT argument that some women are socialized away from technology at an early age. Many of the participants studied non-technology fields in college, but changed their paths once exposed to IT as part of another job. However, the women chosen did not confirm the argument that they were turned away from technology as a result of their college coursework. Out of the 10 women, only one indicated a struggle with college coursework — math and science, in particular. She stated, "My math level is junior college. I don't have the math background to go in and pass that part of the [PMP] test. So, until I go and get that supplemental education, I would never pass that exam."

She was also the only individual in the study without a degree. The remaining nine women had either an associates, bachelors or a master's degree. More importantly, half of the women revealed that that they are under-represented in IT because the field is viewed by most women as a male-dominated field. The women supported comments that suggest that IT organizations are predominantly male which creates a culture hostile to women.

Despite their attempts to influence other women to consider IT as a career, women reflect back of IT leaders as mostly male. This visual imbalance perpetuates the stereotype as a male-dominated versus a "mostly" male field.

In my study, it was noted by one researcher who argued that women are rewarded for showing masculine leadership and influence behaviors, but men are criticized for acting in a feminine manner. For most women, to hear that they 'think like a man' would most likely be considered a compliment, whereas, for a man to hear that he 'thinks like a woman', it would often produce a negative response.

The women in this study defied these comments, as they didn't try to develop masculine styles of communication as a means of fitting into the male-dominated culture. Instead, they insisted on demanding respect from their male colleagues.

As stated by one, "It's just that there's not a lot of women in technology. It is very stressful because as soon as you walk in, men don't see a woman as an IT professional. They don't respect you. I think women go in it and they are ridiculed or put on the project that nobody wants. Or they are put in a support role. And they say, "I'm going to go get myself into another field."

7. Half had negative or challenged relationships with the superiors.

Half of the leaders noted they had strained relationships with their superiors. As illustrated by one, "With my manager, which is the general manager, I feel like it's a little strained right now, but we will get through that. We have two different management styles and I do my best to adapt, but sometimes I forget."

The women also revealed a perception that the 'good ole boy' network still dominates their relationships with their superiors. If they're not a part of the network, their communication with their superiors diminishes to the point of them being a tolerated contributor vs. an asset to the organization. One woman stated of her relationship with her superior, "It is very bad, but it hadn't always been that way. I was a pretty good bragging tool for a minute because I came from higher education. I thought I was a good front man initially."

On this same subject, another woman said, "My direct superiors and I do not get along. Again, older generation, younger generation. There was a distinct class between how I handled my job. I wanted to get there at 7:00 AM. I wanted to work until 5:00 PM. I was not on a retirement status quo. The key five leadership positions [in my organization] will all retire within the next five years. So for them to want to take on something new and something big, something trendy - they were like, 'No way. We're not taking that on because we're almost out,' and I was the girl that wanted to push the envelope. I wanted to do the newest, coolest, best thing and that didn't always work."

Many of the participants experienced negative performance reviews and interpersonal relationships with their male colleagues. Tension with their superiors became the norm. Earlier we stated when discussing the relationship between gender and social desirability, we found that women frequently over-reported favorable behavior on surveys when asked questions about available resources, fairness at work, care and concern, employee trust and company reputation. This suggests that women are more conscious of the impact of a negative rating and tend to compensate by scoring themselves higher as a group than men. However, this aspect was not addressed by this group of women.

8. Most indicated that domain experience was an important factor

in the advancement of women to leadership roles in IT.

Various perceptions were revealed about the skills required to sustain their leadership roles, and no clear answer could be given. However, most important was the 22 elements that emerged from this category about what the women thought of as important factors supporting the advancement of women into IT leadership roles.

The responses given crossed paths frequently, with many citing similar feelings.

22 Elements That Define Important Factors Supporting Advancement of Women in IT Roles

This section proved to be a surprising by-product of this study, showing the elements that support the advancement of women in IT leaderships roles. They are as follows:

1. *Influences to pursue an IT career*

- Because it is male dominated – 1 (10%)
- Exposure as part of a different job – 6 (60%)
- Interest in high school – 4 (40%)

Summary: While most of the women gained interest accidentally while working in other jobs and four others found interest in high school, one stated that her organization was willing to train an IT financial reporter and she just happened to be there. In other words, 7 out of 10 women found IT by accident.

2. *Influences to pursue a leadership role*

- Exposure as part of a different job – 1 (10%)
- Leader by default – 1 (10%)
- Money – 2 (20%)
- Natural progression – 3 (30%)
- Recognition by management – 1 (10%)

Summary: Two women chose an IT career for its monetary benefits, however, it's worth mentioning that out of 10 women, only one held an IT degree with the intent of pursuing an IT career.

3. *Feelings about education and experience preparation*

- Not prepared by education – 7 (70%)
- Not prepared by experience – 5 (50%)
- Prepared by education – 3 (30%)
- Prepared by experience – 5 (50%)

Summary: Thirty percent felt their education was adequate to enter the IT field. One came from a military environment that prepared her technically, while another agreed her education was satisfactory, but that management courses would have prepared her better for an IT leadership position.

4. *Feelings about ideal career path for women*

- Corporate training program first – 1 (10%)
- Entry level experience first in domain –

8 (80%)
- Experience and no degree – 1 (10%)
- Higher education than normal – 1 (10%)

Summary: Most agreed that education held the key to an ideal career path. Two women suggested working as interns to get a sense of the company's culture and/or the preferred area of discipline in the IT industry.

5. *Feelings about career path difference from ideal path*

- Did not take the ideal path – 8 (80%)
- Took the ideal path – 2 (20%)

Summary: Most agreed that they did not follow the ideal career path (lack of the proper education, too much training in teaching, too many years of experience outside of IT). Out of the two women who did, one already held a degree in technology and the other was recruited by being in the right place at the right time.

6. *Feelings about career path difference*

from men

- Different from men – 5 (50%)
- Not different from men – 4 (40%)

Summary: While four women felt their career paths were parallel to men, two of the women specifically mentioned the "good ole boy" syndrome as being obvious within their organization. One woman felt she had to produce at a 190% rate to be equal to the men in her organization.

7. Feelings about representing women

- Did not realize it – 3 (30%)
- Feels good – 2 (20%)
- Haven't thought about it – 2 (20%)
- Humbling – 1 (10%)
- Lonely – 1 (10%)
- Responsibility – 1 (10%)
- Sad – 2 (20%)
- Stressful – 1 (10%)

Summary: Three women felt humbled and honored representing the small force of women serving in IT leadership roles. Two claimed not to be aware of the distinction while three felt sad, lonely or stressed by their roles in representing women.

8. *Feelings about reasons more men are promoted*

- Comfortable working with other men – 7 (70%)
- Male stereotypes – 7 (70%)
- Female stereotypes – 3 (30%)
- Not many women in technology – 2 (20%0
- Because it is acceptable – 1 (10%)
- Hiring a woman requires justification – 1 (10%)
- There is no challenge – 1 (10%)

Summary: Three women felt men were non-emotional which made them better leaders, with one other citing women as being more emotional. One stated that women prioritize duties differently which sets them apart from men, while another stated men are more comfortable dealing with other men. Another said that men don't feel the same pressures as women.

9. *Feelings about competing with men*

- I don't compete – 3 (30%)
- Intimidation – 1 (10%)
- No problem – 1 (10%)
- Not sure – 2 (20%)

Summary: One woman felt you can't compete with men and one refused to compete. Three felt the playing field was equal, and the remaining women did not respond to this element.

10. *Why women are under-represented in IT*

- Male dominated field – 5 (50%)
- Not an appealing field – 3 (30%)
- Lack of math/science education – 3 (30%)
- Lack of interest or exposure to field – 3 (30%)
- Female stereotypes – 2 (20%)
- Lack of leadership background – 1 (10%)

Summary: Five women stated it's a

male-dominated world, with one adding that women in her organization were given projects no one else wanted to work on, lesser support roles or they feared of being ridiculed, while others stated fear of technology (science, math), lack of knowledge, an "unglamorous job". One woman simply didn't really know why women were under-represented.

11. *Sacrifices to obtain and maintain leadership role*

- Family – 4 (40%)
- None – 3 (30%)
- Personal relationships – 3 (30%)
- Reduced day-to-day interaction – 2 (20%)
- Shift to focus on self – 1 (10%)

Summary: Three women felt no sacrifices were made for their careers, while others cited family and relationship compromises of various types. One woman said required traveling sacrificed her personal life.

12. *Feelings about good leadership*

- Ability to communicate – 2 (20%)
- Ability to follow laws – 1 (10%)
- Ability to follow policies – 1 (10%)
- Ability to follow rules – 1 (10%)
- Ability to teach and lead – 1 (10%)
- Business acumen – 2 (20%)
- Creative thinking – 1 (10%)
- Decision making – 1 (10%)
- Dependability – 1 (10%)
- Flexibility – 1 (10%)
- Good work ethic – 1 (10%)
- Honest – 1 (10%)
- Interpersonal skills – 5 (50%)
- Know your team – 3 (30%)
- Knowledgeable in domain – 3 (30%)
- Organized – 1 (10%)
- Respect – 2 (20%)
- Supportive – 7 (70%)
- Unlimited thinking – 1 (10%)
- Willing to learn – 1 (10%)
- Ability to accept criticism – 1 (10%)
- Casual – 1 (10%)
- Hands off – 1 (10%)
- Honesty – 1 (10%)
- Humility – 1 (10%)
- Influential – 1 (10%)
- Open door – 1 (10%)

- Organized – 1 (10%)
- Personable – 1 (10%)
- Results oriented – 2 (20%)

Summary: Being supportive to subordinates was mentioned by most, with interpersonal skills, knowing your team and communicative skills coming in next.

13. *Required skills for women in IT leadership*

- Ability to adapt – 2 (20%)
- Attention to detail – 3 (30%)
- Confidence – 1 (10%)
- Creativity – 1 (10%)
- Dignity – 1 (10%)
- Education – 1 (10%)
- Endurance – 1 (10%)
- Experience in domain – 4 (40%)
- Humility – 1 (10%)
- Interpersonal skills– 3 (30%)
- Manage without emotions – 1 (10%)
- Perseverance – 3 (30%)
- Problem solving – 1 (10%)
- Quick learner – 1 (10%)
- Real life experience – 1 (10%)
- Strong mind – 1 (10%)

- Thick skin – 1 (10%)
- Think around corners – 1 (10%)
- Who you know – 1 (10%)
- Willingness to learn – 1 (10%)

Summary: Two women mentioned needing "thick skin" and remaining unemotional during stressful or conflictive times. Two stated the ability to adapt to situations as they arise, and three mentioned perseverance (and endurance) to "hit the ground running."

14. *Feelings about how women obtain required skills*

- Know your domain – 1 (10%)
- Opportunity – 1 (10%)
- Start early – 2 (20%)
- Through experience – 1 (10%)
- Training – 1 (10%)
- We have it naturally – 4 (40%)

Summary: One woman mentioned training via conflict resolution and additional learning on how to manage teams, while the rest basically agreed that on-the-job experience is the best way to learn IT leadership skills.

15. *Feelings about factors that help women advance*

- Entry level experience first in domain – 4 (40%)
- Confidence – 3 (30%)
- Interpersonal skills – 3 (30%)
- A network – 2 (20%)
- Leadership track record – 2 (20%)
- Education – 2 (20%)
- Adaptability – 1 (10%)
- Leadership changes – 1 (10%)

Summary: Interestingly, experience superseded education, with two mentioning networking skills and adaptability. One thought that not bringing attention to yourself by being mindful of what you wear to work helps advance your career, i.e., blending in with your surroundings.

16. *Feelings about job security*

- Confident – 7 (70%)
- Not confident – 1 (10%)

- Not sure – 2 (20%)

Summary: Job security was strong in spite of many stating their relationships with superiors were strained at best. One stated that job security achievement is attained by "showing your worth", while another said she had job security, but at a cost. She felt overworked and didn't receive enough help when working on projects. "The buck stopped with me."

Another woman commented, "It is not a worry in the sense that it was so political. And it's so all about who you could speak to about what and as long as you were on the right team then you were fine. Education as a whole is very difficult for anything to get fired over...you have to really screw up to lose your job. As long as you wanted to be there, you could be there."

17. *Feelings about factors to influence staying or leaving*

- Career ladder limitation – 1 (10%)
- Changes in job – 2 (20%)
- Changes in leadership – 3 (30%)
- Excuses – 1 (10%)
- Family issues – 1 (10%)
- Increased responsibility – 1 (10%)
- Lack of skills – 1 (10%)
- More benefits – 1 (10%)
- More money – 1 (10%)

Summary: The answers varied with most of the women, citing more personal circumstances other than a change in their leadership hierarchy, which was mentioned by three women. Better opportunities outside of the organization that were more fun or interesting or provided more potential for advancement appealed to two women.

18. *Feelings about factors that hinder women from advancing*

- A negative boss – 1 (10%)
- Age – 1 (10%)
- Children – 1 (10%)
- Dependence – 1 (10%)

- Double standards – 1 (10%)
- Education – 1 (10%)
- Experience level – 1 (10%)
- Inappropriate escalation – 1 (10%)
- Individuality – 1 (10%)
- Lack of role models – 1 (10%)
- Lack of self confidence – 2 (20%)
- Lack of technical skills – 2 (20%)
- Lack of understanding big picture – 1 (10%)
- Life skills – 1 (10%)
- Overstepping one's role – 1 (10%)
- Personal appearance – 1 (10%)
- Poor social skills – 1 (10%)
- Priority on family – 2 (20%)
- Self denial – 1 (10%)

Summary: This category seemed to point to personal situations within the workplace by most women, with lack of skills or lack of confidence and family priorities listed most.

19. *Relationship with colleagues (subordinates, peers and superiors)*

- Peers-positive – 7 (70%)
- Peers-negative – 3 (30%)
- Subordinates – positive – 5 (50%)

- Subordinates – negative – 4 (40%)
- Superiors – positive – 4 (40%)
- Superiors – negative – 6 (60%)

Summary: Most women felt their relationships with their co-workers were solid, with only one stating she didn't believe her colleagues were comfortable talking to her.

20. *Feelings about personal achievements*

- Indifferent – 5 (50%)

Summary: Personal achievements did not weigh heavily on the minds of these women who saw their accomplishments as just doing their job.

One stated that it just wasn't important to her.

21. *Experiences with mentoring*

- None – 3 (30%)
- Some experience – 7 (70%)

Summary: Seven out of the 10 women had strong experiences with mentoring. Nine out of 10 stated they would be more than willing to mentor other women. The remaining woman felt she "wasn't on the same page" as other women to become a mentor them.

22. *Feelings about breaking the glass ceiling*

- Did not know – 2 (20%)
- Doesn't feel like it – 2 (20%)
- Exceed expectations – 2 (20%)
- Followed the rules – 1 (10%)
- It was a natural progression – 3 (30%)
- Listened to mentors – 1 (10%)
- Played the game – 1 (10%)
- Who you know – 1 (10%)
- Doesn't exist – 2 (20%)
- Right place. Right time. – 1 (10%)

Summary: While one found there was no other option for her, most agreed that putting in extra hours, following protocol and taking specialized classes helped them succeed.

Research shows that women are both successful and effective in organizations who have a combination of communication, management and functional skills, networking, but in this group, only 40% held this view. The majority of the women did not have a common set of factors that supported advancement.

Most of the women agreed that domain experience, communication and management skills are important characteristics of women that advance into IT leadership roles. However, the study did not confirm the importance of networking as an important trait of successful women. Although it can be argued that a gender imbalance results in women being frequently marginalized and excluded from social networks predominated by men, the women revealed that either they gave up on accessing the social networks enjoyed by men, or it just simply wasn't a priority for them.

9. The majority said that they were confident about job security.

This revealed an unexpected correlation. Despite half of the study participants stating that challenges existed in their relationship with superiors, seven of the 10 participants expressed confidence in their job security. Of the seven participants expressing this perception, five revealed negative working relationships with their superiors. The results showed that despite the challenges and their uncertain relationships with leaders, the women were still optimistic about being able to sustain their IT positions.

As one woman stated, "Our company is not known for layoffs or termination of people. They really work with their staff to keep their staff members, and that's across the board. So as far as that goes, I'm not fearful that I will be laid off or terminated." She also added that the downside to that is bad managers could maintain their positions within the company for years.

10. The majority said they had some experience with mentoring over the course of their careers.

In a study of six female and six male directors at a leading company in the United Kingdom, it was found that although both men and women face the same barriers to career advancement, men held a distinct advantage over women as a result of challenges women face with work/life balance. Each of the six women in the study not only had mentors, but attributed mentoring as the key to their success. In organizations with an ingrained political culture, women generally faced resistance from men when they attempted to develop political relationships in the organization.

One woman said her mentors – all male – broke the ice in her relationships with other men. "They introduced me to some of the other men on the team so some of that stigmatism like, 'Oh, she's a female developer, a female techie,' so some of that is kind of disseminated because, 'Hey, one of your boys introduced me.'"

Another stated her mentors (also male) were both good and bad: "Learn the behaviors you believe that make them good leaders and look at what you think makes them a good leader. And that's what you want to emulate. The ones that are bad, you want to look at why are they bad and learn from that too."

Women who are mentored have an opportunity to develop skills to overcome or navigate challenging relationships with peers. Here, the women were able to obtain and sustain leadership roles despite the fact that no formal mentoring programs were accessed during their leadership progression.

Seven of the 10 participants had

some experience with mentoring others. However, the manner in which the women described their mentoring experiences also showed possible acquaintances versus developing relationships. The women's response was mixed, however, all agreed, either formal or not, mentoring had a positive influence on women's ability to obtain and sustain their leadership role.

CONCLUSION

The information in this book is significant to the accumulation of knowledge for gender-related issues and organizational studies, and is guided by the research question of *How do women in IT leadership roles perceive how they obtained and sustained their positions?*

Never before has the subject been explored for women in IT leadership roles at such depth. It is my hope that the information regarding the barriers common among women in IT leadership positions might prompt organizations to evaluate their practices and develop or enhance programs that support the career growth of women within IT. An increase in the number of women holding IT leadership positions could offer distinct perspectives on the processes, products, services and strategic direction of organizations that are currently shaped primarily by predominant male perspectives.

A heightened awareness into the understanding of this phenomenon can also help academic institutions and organizations to focus research and programs on why women are under-represented in IT and therefore, develop solutions to this persistent problem.

An increase in the number of women in IT leadership roles can build both the diversity and profitability of organizations, and truly help the world become a better place to live in.

Final Thoughts

This study explored the perceptions and the lived experiences of 10 women that held IT leadership roles for a minimum period of three years at companies in the Dallas/Fort Worth area. Ten key themes emerged that highlighted the influences, barriers, challenges, and the perceptions of women living the experience.

The themes provide a framework

for understanding and improving programs in high schools, institutions for higher education, the military and corporations to increase the level of encouragement and create a culture of gender equality in which women achieve leadership roles proportionate to their numbers in the workforce. It also needs noting that further studies might also show that more women hold leadership positions than current literature available suggests.

The significance of the study is quite important, as so much evidence exists in terms of the benefits of diversity within organizations. In addition, from a practical level, as technology drives the economy and the speed at which innovation is enacted, women need to participate versus accept what men have created. This not only makes the technology more applicable to all, but also opens the door for increased revenue, given the increased relevance of the female demographic. That is only if women contribute to the creation of technology, then therefore the adoption by women may be greater as well.

Given the narrow scope of this study, additional research is needed to validate the key themes that emerged through my interviews with the 10 women who so graciously donated their time, and who are consciously laying down a path of understanding for future generations of IT professionals.

Additionally, understanding the similarities and differences of both male and female groups may help institutions and organizations create programs to increase cross-awareness that this is truly a problem for technology. Formal mentoring programs would also be a valuable tool for organizations to initiate and bring in more women IT professionals.

A final recommendation is to replicate the study with female participants that achieved a leadership role in IT after studying a different discipline in college. This study revealed that 70% of the women had leadership success despite not having formal technology education. It also aided in the development of understanding the similarities or differences between women with an undergraduate technology education versus those that moved into IT from another field.

****

My intention is for IT leaders and educators to put on their thinking caps and find solutions to this global problem. It is also my sincere wish that this study may help organizations expand their programs to include non-technology careers as an alternate path to IT leadership roles, and to spark interest in graduate students with the propensity to continue on with this work. The next phase of my career is dedicated to teaching, writing and speaking on this topic, as well as coaching executives to sustain their roles. To read this dissertation work in its entirety, it can be purchased by contacting me through my email: **drcedricalford@gmail.com**, or you can email me for more information on how to apply the research, thereby

supporting the growth of both young
and mature female IT professionals.

ABOUT THE AUTHOR

Dr. Cedric Alford is an accomplished leader with over 20 years of experience in marketing, technology and business strategy. He has profound experience in designing and executing data-driven marketing strategies for industry leaders across various sectors, including financial services, healthcare, education, professional services and retail.

An innovative and visionary

thinker with a track record of leadership and mentorship, Dr. Alford has held the professional titles of Visiting Professor of Marketing and Management, National Director and Vice President. His current role is a Marketing Solutions Executive for the Microsoft Corporation.

He is a recognized expert in Software as a Service (SaaS) with the communication skills necessary for securing internal support for initiatives, building and energizing high-performing teams, and forging together collaborative partnerships that lead to superior results.

www.ingramcontent.com/pod-product-compliance
Lightning Source LLC
Chambersburg PA
CBHW021118210326
41598CB00017B/1493